EFCE Event No. 320

MULTI-STREAM '85
The Subject Groups Symposium

Organised by The Institution of Chemical Engineers and held at The City University, 16–18th April 1985.

Technical Papers Committee

Professor G.G. Haselden	University of Leeds (Chairman)
A.C. Barrell	HSE (Safety and Loss Prevention Subject Group)
Dr R.J.P. Brierley	ICI (Fluid Separation Processes Subject Group)
Dr P.J. Heggs	University of Leeds (Heat Transfer Subject Group)
Professor C. McGreavy	University of Leeds (Applied Categories and CAPE Subject Group)
Dr J.C. Middleton	ICI (Fluid Mixing Subject Group)
P. Steward	Consultant (Oil and Natural Gas Subject Group)

Subject Group Co-ordinators

M. Blanchard	SACS (Process Control Subject Group)
Dr J.A. Howell	University College Swansea (Biochemical Engineering Subject Group)
Dr D.A. Lihou	Lihou Loss Prevention Service (Safety and Loss Prevention Subject Group)
J.G. Peacock	Protech International (Oil and Natural Gas Subject Group (SONG))
N.R. Twaite	British Sugar Corporation (Solids Drying Subject Group)

**THE INSTITUTION OF CHEMICAL ENGINEERS
SYMPOSIUM SERIES No. 94**

ISBN 0 85295 190 6

Preface

Living systems evolve. This statement is certainly true of Subject Group Symposia,
of which Multistream '85 is only the second offspring.

The main purpose of symposia is the direct exchange of information and ideas,
the value of the associated publications varies depending on the level of sophistic-
ation and consolidation which the topic has achieved. If the concepts are very
new, and the associated measurements or calculations are still exploratory, then
the written material will be ephemeral. This is the nature of the Research Sessions,
and its paper work is included in these pamphlets.

As a concept progresses, and becomes the subject of detailed theory and experi-
ment, it must be described in precise language and symbols, and the resulting
information can have long term value. The formal Technical Sessions aspire to
this level, and it is hoped that the papers in the Proceedings volume prove the point.

The range of subjects covered here respresents a related group of chemical engineering
themes all in the 'mid-stream' of technical interest and achievement. The specialist in
any of the topics covered will undoubtedly be stimulated by what he finds. The non-
specialist could find that browsing is rewarded.

Geoffrey Haselden
(Chairman of Organising Committee)

Contents

* These papers were not available at the time of printing.

THE INSTITUTION OF CHEMICAL ENGINEERS

SOLIDS DRYING GROUP - WEDNESDAY 17 APRIL 1985

INFORMAL SESSIONS - ABSTRACTS OF PRESENTATIONS

(1) A Scientific Approach to Saving Energy in Fluid Bed Driers

by

R E Bahu

SPS Dryer Advisory Service
Harwell Laboratory

Thermal drying processes require a large input of energy because of the unusually large latent heat of vaporisation of water. They accounts for some 12% of the overall energy consumption in such major industrial sectors as food and agriculture, chemicals, textiles, paper, ceramics and timber. The rapid rise in fuel prices over the past decade has stimulated an upsurge of interest in developing improved simulation models for dryers. On existing dryers such models offer considerable potential for saving energy. They can be used to optimise the dryer operation and to evaluate the benefits of hardware modifications such as heat recovery from the exhaust air.

The first step in implementing an energy saving programme on a fluid bed dryer is to perform detailed on-site measurements to establish the material and energy flows into and out of the unit. The results provide firstly, the necessary operating data to input into the simulation model and secondly, a base case against which future improvements can be judged. The measurement techniques involved in the on-site tests will be described briefly.

The fluid bed dryer simulation model developed by SPS consists of two subsidiary models. These are, firstly, a 'material' model describing the relevant characteristics of the material being dried, and secondly, an 'equipment' model describing the relevant characteristics of the equipment in which the drying is being carried out. If the moisture content X of a typical particle in the bed decreases with time t according to a function X(t) and the residence time distribution of the particles is given by a second function E(t), integration over all possible residence times will yield the average product moisture content $\bar{X}o$ as

$$\bar{X}o \;=\; \int_{o}^{\infty} X(t)\;E(t)\;dt$$

This equation describes the performance of a continuous fluid bed dryer. The material model gives X(t) and the equipment model is the function E(t). The material model enables the drying curve X(t) to be calculated at the required bed depth, gas velocity and humidity for the continuous full scale plant from a batch drying test performed in a laboratory unit under any convenient set of conditions.

The equipment model is simply the residence time distribution function E(t). There are essentially three types of fluid bed dryer:

(i)   Single perfectly mixed stage,

(ii)   Several perfectly mixed stages in series,

(iii)   Dispersed plug flow.

Each types has a different E(t) function. The main features of the simulation model will be described.

Finally, a hypothetical example of a plug flow fluid bed dryer will be presented to show how the stimulation model can be used to explore a number of energy saving options.

(2) <u>Use of Fluid Bed Coolers for Moisture Control</u>

by

Dr J Ashworth
Consultant

Drying Research Ltd, Wolverhampton

For many drying processes, substantial sums of money are lost each year due to overdrying in order to meet product moisture specifications. One way of improving product moisture control is to exploit the moisture smoothing capability of the product cooler associated with the drying system.

The dynamics of product coolers are more complex than often realised. In addition to cooling the material, drying and re-wetting of the product can also occur under certain conditions. This is illustrated by a set of performance curves for a fluid bed cooler treating a moderately hygroscopic material. At the beginning of the cycle, the product cools rapidly assisted by an evaporative cooling mechanism due to further drying. As the process continues, the point is eventually reached when the vapour pressure of the moisture remaining in the material is less than that of the cooling air. Hence for the remainder of the cycle the material slowly absorbes moisture.

The balance of the drying/re-wetting process depends upon the initial moisture content of the material. As a net re-wetting effect is observed for dry material and a net drying effect for wet material, the cooler smooths out the product moisture fluctuations. It is further shown that the moisture control effectiveness depends on the cooler air temperature and humidity and the solids to air flow rates. The best control strategy involves maintaining constant air temperature and manipulating the air relative humidity as there is a linear response between relative humidity and final product moisture.

(3) Designing Fluid Bed Dryers for Low Energy Operations

by

H Bender Mortensen et al

A/S Niro Atomizer
Copenhagen, Denmark

The energy required for drying constitutes an essential part of the total energy used for the production of many materials. Therefore, the design of the drying stage must be done with great care.

Fluidized beds are widely applied in industry for processes such as drying, cooling and chemical reaction. Fluidized bed dryers are used for e.g. polymers, organic and inorganic products, pharmaceuticals and dairy products. The fluidized bed dryers can handle raw materials in a broad range of physical conditions from free-flowing solids through filter or centrifuge cakes to solutions, melts or suspensions.

Fluid bed dryers with very different characteristics are in operation e.g. stationery or vibrating units, circular or rectangular configurations comprising single or multi stage designs. Furthermore operational variations are available with respect to constant or pulsating gas flows, the provision of mechanical agitation and/or the installation of heating surfaces within the bed. The solids flow pattern in continuous units may vary from fully back mix to plug flow and from cross flow to counter current flow relative to the flow of the fluidizing gas.

The objective of this paper is to demonstrate how the optimization of heat economy of industrial scale fluidized bed drying systems may be accomplished by selection of:-

- Solids flow pattern (back-mix versus varying degrees of plug flow)

- Gas/solids flow pattern (single or multi-deck fluid beds)

- Contact heating elements

- Heat recovery

- Operating parameters

In order to illustrate the above factors, the paper focuses on fluidized bed drying systems for hygroscopic as well as non-hygroscopic materials. Hexane, wet polypropylene (hygroscopic) and water-wet suspension PVC (non-hygroscopic) are used as examples.

The study shows that significant improvements in the operating costs of fluidized bed drying systems may be achieved by analysing the drying systems with regard to a number of process parameters and fluidized bed design characteristics. In principle, similar evaluations may be used for other products based upon their specific drying characteristics.

(4) **The Design of Distributors for Gas Fluidized Beds**

by

Dr D Geldart
Chairman

Postgraduate School of Studies in Powder Technology
University of Bradford

The performance of the gas distributor often determines the success or failure of a fluidized bed dryer and although much more is known now than 20 years ago, there are still many pitfalls for the designer.

Particle and gas properties play a key role in the successful design together with the critical pressure drop ratio, and hole size, geometry and spacing; these strongly influence jet penetration, dead zones, particle sifting, attrition and mixing.

The current state of the art is reviewed in the light of recent research and industrial experience.

(5) **The Improved Batch Fluid Bed Process**

by

K Huddlestone
Managing Director

ACM Machinery Ltd, Basingstoke

The improved batch fluid bed process will be discussed. A review will be presented of some recent advances in respect of the designs control and hazard containment of fluid bed driers.

(6) <u>Fluid Bed Drying and Cooling in the Sugar Industry</u>

by

N R Twaite
Chief Process & Design Engineer

British Sugar plc

Traditional methods of drying both white and brown sugar crystals are given. The problems with water removal from the crystal face and the influence of impurities affecting the hygroscopic nature of the film are discussed.

The design parameters of the crystal/air system are examined and figures for use in the drying operation established.

The development of fluid beds on crystalline sucrose is discussed with some performance figures and the pilot plant and full scale trials in British Sugar cited. Problems in the use on brown sugars are highlighted and potential solutions discussed.

A HIGH PRESSURE MICROREACTOR FOR CATALYTIC HYDROPROCESSING OF COAL DERIVED LIQUIDS

R.I.Arnott* and P.Sunderland*

A multi-mode microreactor system has been constructed to study catalytic hydrotreating of oxygen, sulphur and nitrogen containing organics typically found in coal derived liquids. The system is designed to operate at temperatures up to 725K and pressures up to 2×10^7 Nm^{-2} using liquid flowrates in the range 1 to 600 ml hr^{-1}. It can work in pulse flow, steady state flow or recycle mode to allow a full investigation of the reaction to be made. A description of the system is given and its successful commissioning demonstrated by means of a series of experiments using furan and tetrahydro-furan.

INTRODUCTION

Oxygen, sulphur and nitrogen compounds occur naturally in coal. When coal is liquefied it is usually done in two stages, a solvation stage and a hydro-processing stage. The feed to the second stage contains the bulk of the organic oxygen, sulphur and nitrogen as relatively small cyclic aromatics similar to those used in this study. The way in which these compounds react in the hydroprocessor has an important bearing on the nature of the products and hence their usefulness or otherwise as fuels or feedstock for the organic chemicals industry. Sulphur and to a lesser extent nitrogen compounds have been well studied but the oxygen containing compounds less so.

In this work the conversion, in the H-donor solvent tetralin, of furan and tetra-hydrofuran, simple analogues of compounds known to be present in real coal derived liquids, has been studied. The catalyst used was Cyanamid HDS2 a commercially available alumina supported cobalt molybdate catalyst. The multi-purpose microreactor designed to carry out the work has been built with low cost, versatility and safety very much in mind. In preliminary work with the reactor the decision was taken to concentrate on a proper commissioning of the apparatus with simple reactants. Subsequently experimentation with mixtures and real coal derived liquids will be carried out.

EXPERIMENTAL

The Reactor Loop (Figure 1)

In view of the hazardous nature of work with hydrogen at high pressure and temperature and the cost of solvent, safety and economy were important considerations when the system was being designed. As a result, the apparatus

* Department of Chemical Engineering, University of Leeds, Leeds LS2 9JT.

was sited in a reinforced concrete autoclave bay.   The rig was constructed from stainless steel throughout and designed to handle pressures up to 2×10^7 Nm^{-2} well beyond the range to be used in the experimental programme.   The system was kept small, the reactor loop containing just 10 ml. of liquid. Flowrates of fresh solvent or solution to the loop were to be no greater than 10 ml.min^{-1}.   The entire apparatus was enclosed in a steel box flushed continuously with nitrogen under positive pressure relative to the atmosphere, venting outside the building.   Additional safety features were built in so that should a leak be detected, water fail, or temperature or pressure become excessive then the rig would automatically shut down.   The control panel was placed in an adjacent room with contact to the apparatus via small holes drilled in the wall. Operators only need enter the autoclave bay when the apparatus is not working. Filtered solvent or premixed solution is drawn from the feed tank through one loop of the pulse valve by means of the feed pump before delivery to the reactor loop.   In pulse mode, solvent is in the feed tank and a known quantity of solution in a second loop of the pulse valve. At the appropriate time the pulse can be automatically injected into the solvent stream.   In steady state or recycle mode the feed tank contains premixed solution.   The feed pump is a standard HPLC pump type 300-01 supplied by A.C.S.Ltd. designed to provide pulse free flow between 0.2 to 10 ml.min^{-1} at pressures of up to 4×10^7 Nm^{-2}.

Feed then passes to the hydrogen saturator where the required degree of saturation can be achieved by bubbling hydrogen through the liquid at a suitable rate.   Introducing the hydrogen at this stage rather than in the reactor itself has the advantage of reducing the number of phases present to just two so that one at least of the transport steps can be discounted when experiments aimed at producing kinetics are carried out. Preheat to reactor temperature occurs in the preheat vessel.

The reactor is a stainless steel tube 15 cm. long with an internal diameter of 6 mm.   It contains the catalyst, particles size range .21 - .42 mm. diameter, and inert packing held in position by small pieces of gauze which also prevent solid particles from passing downstream of the reactor.   The tube is mounted along the axis of a copper cylinder 10 cm. in diameter sufficiently massive to ensure even temperature distribution in the central part of the tube.   Thermocouples measure the reactor temperature and serve as probes for the temperature controller.   All temperatures measured including both preheat and reactor can be displayed on the control panel.   Heat input is via a heating tape wound around the outside of the copper block.   The whole is adequately insulated from the surroundings by means of vermiculite beads. This set up is capable of operating at 800K for long periods without failture although in practice it is called upon to operate at no more than 623K.   Pressure before and after the reactor is measured using pressure transducers and displayed on the control panel.

After leaving the reactor, the products are cooled in a water cooled quench and then either expanded in a controlled fashion to atmospheric pressure through the fine control pressure reduction valve or alternatively pass to the recycle pump.   This is a duplex pump type Q1/SS10C supplied by Metering Pumps Ltd.   It has a flame-proofed motor and can deliver up to 160 ml.min^{-1} at 2.4×10^7 Nm^{-2}.

Effluent leaving the pressure reduction valve passes through a sample valve and then either to a gas chromatograph for analysis or to the product tank.

<u>The Hydrogen Feed System</u> (Figure 2)

Hydrogen is fed to the hydrogen pump from a hydrogen cylinder after the pressure has been reduced from full cylinder pressure to about 10^6 Nm^{-2}. The hydrogen pump or intensifier is a Haskel air driven gas booster type ACT30/75. It serves to top up and maintain automatically at some preselected pressure the high pressure storage vessel which is as the name implies the source of hydrogen for the reactor. The vessel is a Baskerville & Lindsay stainless steel autoclave type of 1 litre capacity, capable of withstanding pressures of 4×10^7 Nm^{-2}. Hydrogen would normally be stored in the vessel at 3×10^7 Nm^{-2}. The hydrogen cylinder and high pressure storage vessel are sited outside the building. There is provision for both manual and automatic shut off in the event of failure so that the hydrogen reservoirs cannot discharge into working areas.

The pressure in the large pressure storage vessel is then reduced to some more suitable upstream pressure via a pressure regulator. Hydrogen at this point can be fed to one of two hydroprocessing rigs. Flow control is achieved by means of a combination of a fine control needle valve and a set of capillaries of varying lengths. The capillaries are connected across an air operated differential pressure meter type 13HA M31 manufactured by Foxboro and capable of measuring Δp in the range 5-62 kPa.

Finally hydrogen passes through a non-return valve before entering the reactor loop at the hydrogen saturator.

The whole is constructed inside a stout steel box and flushed continuously by air drawn through the box by means of a flame-proofed extractor fan and then vented safely outside the building.

Analysis

Feed and effluent samples are anlaysed using a Pye 204 gas chromatograph fitted with a thermal conductivity detector. A 1.5 m column containing apiezon-L on celite at 423K was used to separate the hydrocarbons and a 1.2 m porapak Q column at 75°C used to separate the oxygen containing compounds. In each case the carrier gas was helium flowing at 45 ml. min^{-1}.

Results

The apparatus has been commissioned using furan and tetrahydrofuran, simple analogues of oxygen containing compounds typically found in coal derived liquids. A 1 weight % solution of each in the H-donor solvent tetralin was passed at flow-rates in the range 0.2 to 2 ml/m over the non-sulphided cobalt molybdate catalyst Cyanamid HDS 2 at pressures between 0.8 and 1.2×10^7 Nm^{-2} within the temperature range 623K to 673K.

Initially work has been carried out in the absence of hydrogen gas because the solution was sufficiently dilute for there to be enough hydrogen available from the donor solvent to bring about measurable reaction. This safer method of operation provided much useful information with regard to the nature of the products, likely reaction schemes and what the catalyst activity was and how it changed qualitatively for each compound studied. The disadvantages of the method were that conversions had of necessity to be kept low, catalyst deactivation was accelerated with some coking occurring at the higher temperatures and naphthalene building up in the solution. It is not possible either to obtain any meaningful kinetic data.

The apparatus was operated in both pulse and steady state mode. The advantage of the former method of operation was in terms of economy because only small quantities of the oxygen containing compounds were required and solvent recovery was possible.   Much useful information was obtained using the pulse technique particularly in the early stages when analytical techniques were being developed but the work was hampered because there was no means of monitoring effluent composition continuously.   The advantages of the steady state mode of operation were better control of pressure, larger samples for analysis and the ability to study catalyst deactivation more effectively as well as producing some non-quantitative kinetic information.

No catalysis of the reaction by the walls of the reactor was observed within the temperature range studied either with solvent alone or with solutions of the oxygen compounds.   There was some thermal decomposition of the solvent at temperatures in excess of 673K but this was outside the range of these experiments.   Neither was any reaction observed when reactants were passed through the reactor packed with catalyst support.   With catalyst, reaction did occur and at temperatures as low as 623K much lower than those at which thermal decomposition of solvent occurred.   The results indicated that reaction schemes similar to those commonly accepted for hydroprocessing the corresponding sulphur compounds hold true for furan and tetrahydrofuran (Figure 3).   There were differences in activity as one would expect but it has not been possible to put these observations on a quantitative footing as yet.

Future Programme

The next phase of the project will be to do a similar range of experiments with hydrogen and then sulphided catalyst with and without hydrogen.   In view of the detail provided by the experiments without hydrogen, only a relatively few experiments with hydrogen should be necessary to provide a complete kinetic picture.   Experiments will be carried out in recycle mode so as to provide information for higher degrees of conversion.   Good recycle rate together with the small catalyst particle size used should ensure that transport effects are minimised.

Of the three possible hydroprocessing reactions that can occur, i.e. hydrocracking, hydrogenolysis and hydrogenation, only the latter two were observed with the compounds used here.   In order to be able to understand the whole picture so that catalyst modifications can be proposed it would be better if oxygen containing compounds were used in which all three occurred.   For this reason, the range of compounds will be extended to include dibenzofuran and substituted furans.   The results should allow predictions to be made for real coal liquids although the intention is eventually to work with such liquids.

CONCLUSIONS

A high pressure micro-reactor system to be used for a study of the catalytic hydroprocessing of synthetic coal derived liquids has been successfully commissioned and a programme of experiments using oxygen containing compounds initiated.

ACKNOWLEDGEMENTS

The authors are grateful for support from the Science Research Council.   We also wish to acknowledge the loan of much of the hydrogen feed system by the N.C.B.(C.R.E.) Stoke Orchard.   Catalyst was supplied by Cyanamid B.V. Rotterdam.

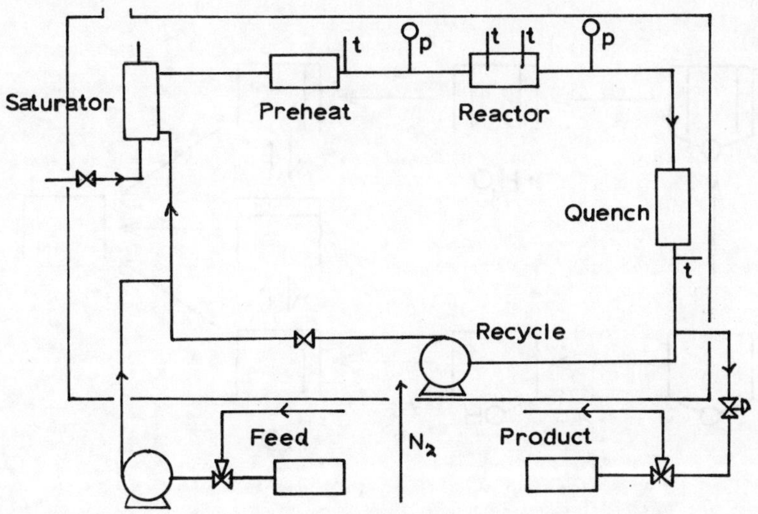

Figure 1.   Reactor loop.

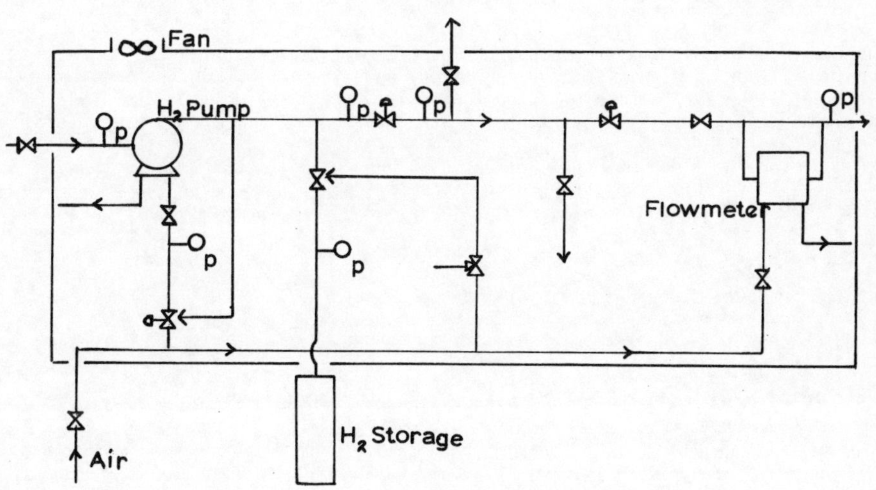

Figure 2.   Hydrogen feed.

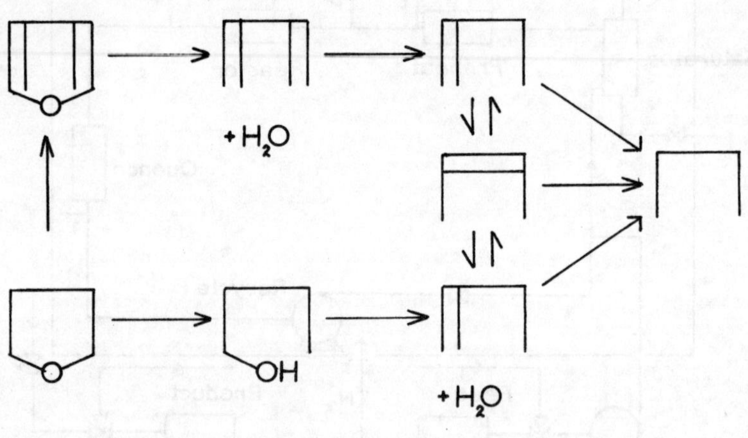

Figure 3.   Reaction of furan and tetrahydrofuran.

THE MECHANISM OF METHANOL SYNTHESIS ON COPPER-ZINC-ALUMINA CATALYSTS

G.D.Short*

The mechanism of methanol synthesis from carbon oxides and hydrogen under commercial conditions on the surface of a reduced copper-zinc-alumina catalyst is discussed. The immediate molecular precursor of methanol is shown to be carbon dioxide, which is replenished via oxidation of carbon monoxide on oxidised surface copper sites as it is consumed. The activity of the catalyst is proportional to the copper metal surface area, and the shift reaction is shown to occur on sites differing from those on which methanol synthesis takes place.

INTRODUCTION

Methanol is synthesised commercially from carbon monoxide, carbon dioxide and hydrogen at temperatures between 220°C and 300°C and at pressures around 50-100 bar. The most common catalyst for this reaction consists of pellets containing alumina, zinc oxide and copper, charged to the reactor as oxidic material and activated prior to use by reduction with hydrogen[1]. Thermodynamically, methanol is not a favoured molecule of those obtainable from synthesis gas reactions and the fact that the selectivity to methanol over $Cu-ZnO-Al_2O_3$ catalysts is well in excess of 99% has for many years stimulated research into the mechanism of methanol synthesis, aided also by a belief that deeper understanding of the mechanism may in turn lead to the design of even more effective catalysts. In common with other catalytic reactions a coherent picture of mechanism does not emerge from the application of one or two investigative techniques, but rather from a broad range of methods. Accordingly this paper presents some of the more significant results obtained over several years in ICI using a variety of techniques, but concentrating where possible on the industrial catalyst exposed to conditions approximate as closely as possible to those encountered in full scale plants[2]. Much of the confusion and controversy in the literature stems from the fact that the mechanism of synthesis (and indeed the nature of the catalyst) is dependent on the reaction conditions being used.

* ICI Agricultural Division, Billingham, Cleveland

EXPERIMENTAL

Catalyst activity is measured under standard conditions of 250°C, 50 bar, space velocity 4×10^4 hrs^{-1} and gas composition 10% CO, 3% CO_2, 67% H_2, 20% N_2 in pseudoisothermal reactors. So long as conversions are small and equilibrium is not too closely approached, activity is proportional to exit methanol concentration. Copper surface area is measured by surface reaction with nitrous oxide under carefully controlled conditions[2]. Temperature programmed desorption and reaction spectroscopy have been well described in other publications[3]. Measurements of radioactivity used standard procedures.

ROLE OF CARBON DIOXIDE

Kagan et al[4] have published evidence that the precursor to methanol is carbon dioxide, while others have refuted the claim. We have extended that work and confirmed that carbon dioxide is indeed the immediate molecular precursor to methanol. With $^{14}CO_2$ as a component, gas containing equal proportions of CO and CO_2 was reacted over $Cu-ZnO-Al_2O_3$ catalyst as described above. Space velocity was varied between 10^4 and 2.4×10^5 hrs^{-1} and exit gases analysed and their radioactivities measured. Results in Figure 1 show clearly that synthesis proceeds via carbon dioxide since scrambling of radioactivity via the shift reaction is negligible at high space velocities, where methanol radioactivity approximates that of the inlet carbon dioxide. It may also be deduced that the site on which shift occurs cannot involve the formate ion, since methanol synthesis is fast whereas reverse shift is slow and if there were a common intermediate the route would be open for methanol synthesis from carbon monoxide.

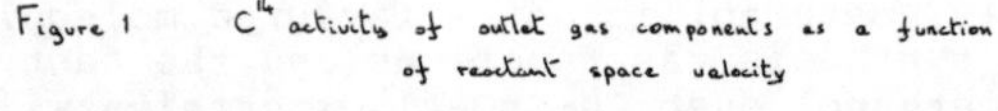

Figure 1 C^{14} activity of outlet gas components as a function of reactant space velocity

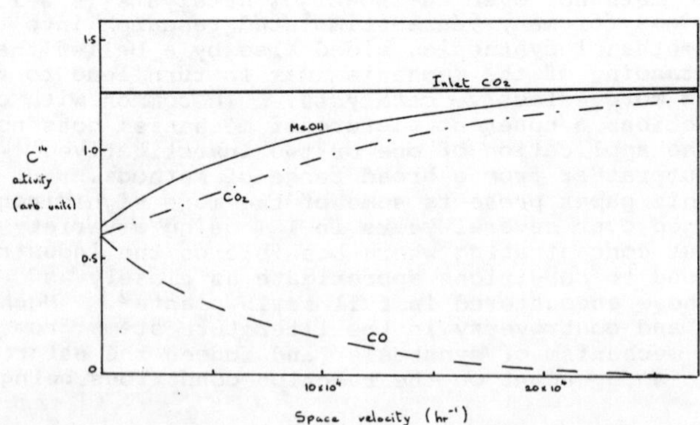

STATE OF COPPER IN A WORKING CATALYST

When different $Cu-ZnO-Al_2O_3$ catalysts are reduced it is found
that their initial copper surface areas are proportional to
their methanol synthesis activity, Figure (2).   Furthermore,
when methanol synthesis is interrupted and surface area re-
measured, it is found that a substantial proportion of the metal
surface is in an oxidised state, but readily reducible to the
original metal surface area by further hydrogen reduction.
Thus in the working state the catalyst surface is partially
oxidised, an observation of considerable significance for the
mechanism of synthesis.   The area measurements strongly suggest
that only copper metal/copper oxide is implicated in the rate
determining step of synthesis.   Other oxides probably
contribute to promotion of initial reduction, and stabilisation
of copper area.

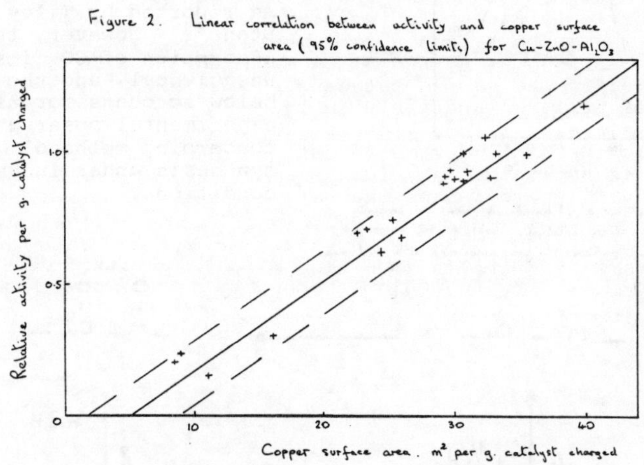

Figure 2.   Linear correlation between activity and copper surface area (95% confidence limits) for $Cu-ZnO-Al_2O_3$

MECHANISM OF METHANOL SYNTHESIS

The combination of temperature programmed desorption (TPD) and
temperature programmed reaction spectroscopy (TPRS) has been
used to examine the mechanism via adsorption, desorption and
decomposition of reaction intermediates.   These techniques have
shown that on $Cu/ZnO/Al_2O_3$ catalysts, the observed intermediate
on the surface of the Cu and ZnO components of the catalyst is
the formate species.   Methanol adsorption (Figure 3) at room
temperature onto a catalyst in which the surface copper is about
30% oxidised (see above) was characterised by two main peaks in
the desorption spectrum: (i) by the coincident desorption of H_2
and CO_2 at a peak maximum temperature of 170°C - a fingerprint
of the existence of a formate species adsorbed on the copper
component[3] and (ii) by the coincident desorption of H_2 and CO at
310°C - characteristic of a formate species adsorbed on zinc
oxide[3,5].

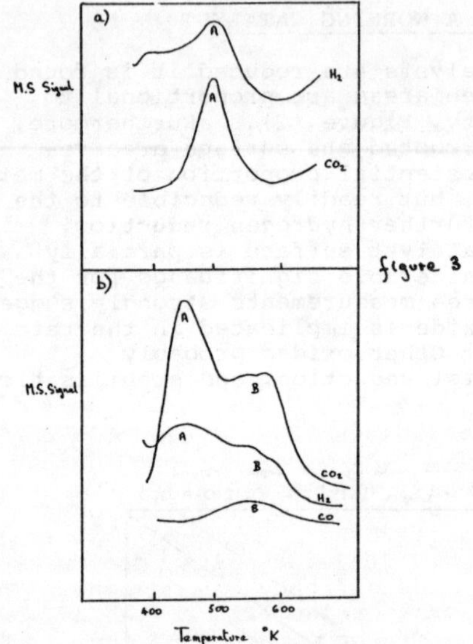

Methanol decomposition on a) unreduced and b) partially reduced Cu – ZnO – Al₂O₃ catalyst. Peaks A are associated with Cu(I) formate. Peaks B are associated with Zn formate.

Adsorption of CO_2 is not facile on copper or oxidised copper at the relatively low pressures accessible by TPD or TPRS. Hence it is not possible unequivocally to distinguish between CO_2 adsorbed by the basic oxide support and its interface with copper, and CO_2 adsorbed at 50-100 ats on partially oxidised copper, as reported by Tiley and Stone[6]. However, the adsorption of CO_2 itself is unequivocal, and the scheme below accounts for all the experimental observations concerning methanol synthesis under industrial conditions.

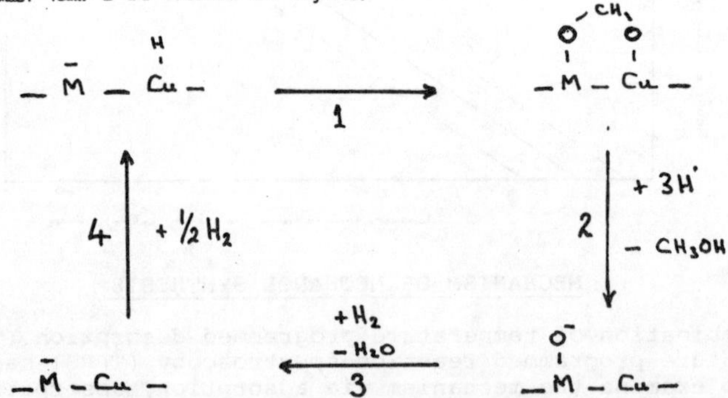

The mechanism does not specify which of the three hydrogenation steps represented by reaction 2 is rate determining, but it seems likely that it is the step which involves carbon-oxygen bond breakage, ie hydrogenolysis.

However, as reaction 2 shows, each copper site at which methanol is synthesised will become oxidised as a consequence of the synthesis reaction. These oxidised sites constitute in the steady state some 30% of the total surface available initially following reduction, the steady state being maintained by the reactions of CO and H_2 with $O_{(ads)}$.

Oxidation of CO on partially oxidised copper constitutes part of the shift reaction, while TPRS shows that CO is not a product of Cu(I) formate decomposition.   The shift reaction, therefore, under methanol synthesis conditions on $Cu-ZnO-Al_2O_3$ catalysts, does not involve the formate intermediate, in support of the deduction from the radioactivity results.

Thus:

$$CO_2 + H_2 \rightleftharpoons X \rightleftharpoons CH_3OH$$
$$CO + H_2O$$

CONCLUSION

Under industrial conditions methanol is synthesised from CO_2 on a partially oxidised copper surface.   There may be a promoting role played by support oxides at the copper-metal interface. The shift reaction, which replenishes CO_2 by CO oxidation on the partially oxidised surface, does not involve a common intermediate with methanol synthesis.   Reverse shift is slow compared with methanol synthesis.   The efficacy and mechanism of synthesis will depend greatly on the composition of the reacting gases, since this will determine the nature of the catalyst surface.

ACKNOWLEDGEMENTS

Discussion with and results obtained by the following colleagues is gratefully acknowledged: Mr G C Chinchen, Dr P J Denny, Dr M S Spencer, (ICI Agricultural Division); Dr D G Parker (ICI Petrochemicals and Plastics Division); Dr K C Waugh (ICI New Science Group), and Dr D A Whan (Edinburgh University).

REFERENCES

1   UK Patent 1159035 (1965).

2   G. C. Chinchen, P. J. Denny, D. G. Parker, G. D. Short, M. S. Spencer, K. C. Waugh and D. A. Whan, Preprints ACS Division of Fuel Chemistry, <u>29</u> (5), 178-88 (1984).

3   M. Bowker, H. Houghton and K. C. Waugh, J. Chem. Soc., Faraday Trans. I, <u>77</u>, 3023 (1981).

4   A. Ya Rozovskii, G. Lin, L. B. Liberov, E. V. Slivinskii, S. M. Loktev, Yu B. Kagan and A. N. Bashkirov., Kin i Kat <u>18</u> (3), 691, (1977).

5   M. Bowker, H. Houghton and K. C. Waugh, J. Chem. Soc., Faraday Trans. I, <u>78</u>, 2573 (1982).

6   R. M. Dell, F. S. Stone and P. F. Tiley, Trans. Farad. Soc., <u>49</u>, 195 (1953).

THE DYNAMICS OF CATALYTIC OXIDATION REACTIONS

S.C. Capsaskis and C.N. Kenney*

Elementary step analysis in which the kinetics of adsorption and desorption steps are also included with surface reaction processes provides a semiquantitative agreement with experimental observations on the catalytic oxidation of olefins over platinum. These include cycling the input concentration; the occurence of self-sustaining oscillations, and period doubling. This latter phenomenon is known to be theoretically possible in oscillatory systems described by non-linear differential equations but has not been described previously in a chemically reacting system.

INTRODUCTION

The study of periodic feed switching in catalytic reactors is a specific example of the wider topic of 'forcing' a system, mechanical, chemical, electrical or environmental, which may have a natural frequency of its own. Work to date has been mainly concerned with devising optimum switching schemes which could enhance reactor performance, as measured in terms of time-averaged conversion or reaction selectivity.

A plethora of reactions have been examined under conditions where the reactor feed concentration varies with time. These include the oxidation of sulphur dioxide on V_2O_5, the oxidation of ethene on Ag, the hydrogenation of butadiene on Ni, the synthesis of ammonia, the addition of ethene to acetic acid catalysed by H_2SO_4, the Fischer-Tropsch synthesis reaction and, last but not least, the oxidation of carbon monoxide. Hawkins [1] and Cutlip [2] reported significantly increased reaction rates of CO oxidation on Pt compared to the steady state rate, when two streams consisting of CO/argon and O_2/argon mixtures are alternately fed to the reactor. The oxidation of CO on V_2O_5 has also been examined as well as the oxidation of CO/NO mixtures on Pt.

The work reported here is concerned with the more fundamental problem of the interaction of a time-varying reactor feed with the complex kinetics of the heterogeneous catalytic oxidation of a CO/propene mixture, a system which can behave as a chemical oscillator [3]. So far, most analytical and numerical work on periodic forcing of chemical oscillators has concentrated on forcing of the Brusselator [4,5].

Bailey [6,7] has established the terms of reference for theoretical discussion of a chemical reaction occuring under oscillatory feed conditions. For the lumped parameter system denoted by Bailey by

$$dx/dt = f(x(t), u(t))$$

*Department of Chemical Engineering, Pembroke Street, Cambridge, CB2 3RA with u(t) representing the τ-periodic forcing term

$$u(t) = u(t+\tau)$$

comparison of the characteristic response time of the system τ_c (as determined by linearisation in the neighbourhood of the steady state(s) of the system) with the period of the applied feed oscillation τ establishes two limiting types of periodic output and an intermediate region where more complicated phenomena can occur. These are:

(1) the quasi-steady state region for wich $\tau \gg \tau_c$. In this region, for the simple case of a periodic square wave input in feed concentration, the system has enough time to recover from each change in the feed composition and attain a steady state appropriate to the feed at that part of the cycle. The dynamics of the system then can be approximated by:

$$x(t) = h(u(t))$$

where h is defined such that

$$0 = f(h(u(t)), u(t))$$

(2) the relaxed steady-state region for which $\tau \ll \tau_c$. Here the dynamic behaviour of the system will be characterised by small amplitude oscillations of period τ near the relaxed steady state defined by:

$$0 = (1/\tau) \,_o\!\int^\tau f(x_{rs}, u(z)) \, dz$$

(3) the intermediate frequency region, examined by Sincic and Bailey [8], where certain behaviour patterns exist which cannot be predicted from the above two asymptotic approximations. Simulations of complex chemical systems were considered to demonstrate that an oscillatory system response can be obtained with a period which is an integral multiple of the forcing period τ (sub-harmonic response), a phenomenon well known in the treatment of non-linear oscillations [9] but not previously described in the catalytic reaction literature.

This paper describes experimental observations of sub-harmonic response (S.H.R.) during the oxidation of CO/C_3H_6 mixtures over supported Pt catalyst in a well mixed flow system where the feed composition of C_3H_6 is cycled with a square-wave feed input. Furthermore it is shown that the elementary step model proposed for CO/alkene oxidation over Pt [10] can account for such period lengthening of the output oscillations.

<u>EXPERIMENTAL</u>

The reactor consists of an aluminum cylinder with seven parallel channels which contain a catalyst consisting of 10.04 g of 0.5% w/w Pt on Al_2O_3, in the form of cylindrical pellets of 1/8 in. diameter (provided by Johnson Matthey Ltd.). A metal bellows pump provides recirculation of the gas stream through the reactor. This arrangement, which is housed inside an oven, has been shown by residence time studies to approximate to a CSTR of free volume 56.6 cc. The pressure in the reactor is 78-87 cm Hg abs. and the temperature range employed was 100-160 deg. C. The mole fraction of up to four gas-phase species (typically CO, O_2, C_3H_6 and CO_2) in the reactor inlet and outlet is monitored at ca. 3 seconds intervals using a VG Micromass-6 magnetic deflection mass spectrometer. Data from the mass spectrometer are recorded by a PDP 11/45 minicomputer for subsequent processing. The experiments described below involved first pretreating the catalyst (usually with a mixture of either 2% CO or 3% O_2 in Ar) followed by manual switching of the reactor feed between FS1 and FS2 at intervals equal to one-half the desired forcing period, so that a square wave oscillatory input concentration was directed to the reactor. The switching pattern was also recorded by the computer for subsequent period verification.

The investigation of reactor response to periodic switching was carried out with three feed mixtures, Scheme X1: Switching between 0.5% C_3H_6, 4% CO, 3% O_2 in Ar and 1% C_3H_6, 4% CO, 3% O_2 in Ar. Scheme X2: Switching between 0.5% C_3H_6, 2% CO, 3% O_2 in Ar and 0.5% C_3H_6, 4% CO, 3% O_2 in Ar. Scheme X3: Switching between 0.5% C_3H_6, 2% CO, 3% O_2 in Ar and 1% C_3H_6, 4% CO, 3% O_2 in Ar.

All experiments were carried out at either 135°C or 150°C (410 K or 423 K).

Scheme X1, in which the two streams differ only in the concentration of propene, was applied at 410 K and with a feed flowrate of 22 cc min^{-1} NTP. Under these conditions, the mixture composed of 1% C_3H_6, 4% CO, 3% O_2 exhibits spontaneous oscillations (3,11) (fig.1) which are also similar to those in which the olefin component is 1-butene (10). When periodic feed switching is applied with a forcing period for example of 50 s (fig.2), the reactor output response consists of concentration oscillations of period equal to the forcing period. The duration of two feed cycles is shown in the figure to provide comparison between feed and the output response.

When the forcing period is reduced to 30 s, the reactor, after about 3 cycles, settles down to a sub-harmonic oscillatory response (SHR) of 60 s period (fig.3) with a characteristic indentation half-way along the decreasing part of the CO_2 trace. A further reduction of the forcing period to 20 s eliminates the reactor oscillations virtually completely.

Mathematical model

The reaction of mixtures of CO, O_2 and C_3H_6 over supported Pt catalysts can be represented in terms of elementary step model similar to the one considered for the reaction of CO, O_2 and 1-C_4H_8 (10) assuming (1) all gases compete for the same sites, (2) oxygen adsorbs on Pt dissociatively, (3) carbon monoxide adsorbs on Pt in the linear form, (4) reaction occurs only between adsorbed species, (5) propene is completely converted to carbon dioxide and water (as suggested in (12)). Steps involving the adsorption and oxidation of propene on the Pt surface

$$C_3H_6 + 2s \rightleftharpoons C_3H_6 - 2s \tag{1}$$

$$C_3H_6 - 2s + 90 - s \rightarrow 3CO_2 + 3H_2O + 11s \tag{2}$$

are added to the basic elementary step model for CO oxidation:

$$CO + s \rightleftharpoons CO\text{-}s \tag{3}$$

$$O_2 + 2s \rightleftharpoons 2\ O\text{-}s \tag{4}$$

$$CO\text{-}s + O\text{-}s \rightarrow CO_2 + 2s \tag{5}$$

In addition, the reversible adsorption of propene and CO_2 on the support material sites, s', is included (rates R_6, R_7) since such an effect is observed experimentally. The reaction steps (1)-(5) can be translated into a set of ordinary differential equations describing the evolution of the reaction in a gradientless reactor by setting down a mass balance for each gas phase and surface species, x_i and z_i respectively. Related reaction terms R_i are formulated in terms of gas phase and surface concentration.

$$dx_1/dt = (x_{1F} - x_1)/\tau - R_1 - x_1\Phi$$

$$dx_2/dt = (x_{2F} - x_2)/\tau - R_2 - x_2\Phi$$

$$dx_3/dt = (x_{3F} - x_3)/\tau - R_4 - R_6 - x_3\Phi$$

$$dx_4/dt = (x_{4F} - x_4)/\tau + R_3 + 3R_5 - R_7 - x_4\Phi$$

$$dz_1/dt = R_1 - R_3$$

Reaction terms

$$R_1 = kp_1 x_1 c_{TOT} z_v - km_1 z_1$$

$$R_2 = kp_2 x_2 c_{TOT} z_v^2 - km_2 z_2^2$$

$$R_3 = k_3 z_1 z_2$$

$$R_4 = kp_4 x_3 c_{TOT} z_v^2 - km_4 z_3$$

$$dz_2/dt = 2R_2 - R_3 - 9R_5 \qquad\qquad R_5 = k_5 z_3 z_2$$

$$dz_3/dt = R_4 - R_5 \qquad\qquad R_6 = kp_6 x_2 c_{TOT} z'_v - km_6 z'_3$$

$$dz'_3/dt = R_6 \qquad\qquad R_7 = kp_7 x_4 c_{TOT} z'_v - km_7 z'_4$$

$$dz'_4/dt = R_7$$

where

$$z_o = (m\ s_o)/(V\ c_{TOT})$$

$$z_{sup} = (m\ s_{sup})/(V\ c_{TOT})$$

$$z_v = z_o - \Sigma z'_i$$

and Φ is a correction term allowing for volume changes during reaction and can
be derived from the condition that $\Sigma(dx_i/dt)=0$. A more detailed description of
the model is given in (10) and has sufficient generality to be able to account,
in semiquantitative terms, for multiple states, hysteresis, relaxation, sinu-
soidal and multipeak oscillations in four different oxygen/carbon monoxide/
alkene reactions over platinum catalysts. Following the nomenclature for homo-
geneous oscillating reactors, it is perhaps appropriate to describe this chemical oscillator model as the 'Cantabrator'.

Simulation

A simpler version of this model, based on equations (3)-(5) has been used
to model forced switching experiments involving only CO and O_2 (13). Such
simulations require values of the various rate constants in the above kinetic
equations (and those derived from separate experiments not involving feed oscillations were used). A comprehensive set of rate constants and the surface
capacity of the catalyst used have been found by us for the carbon monoxide/
propene oxidation system at 423 K and these were employed, even though the observed occurences of S.H.R. were at 410 K.

Periodic feed switching experiments were simulated by integrating the
ordinary differential equations presented in the previous section using a variable order, variable step Gear method, available on the Cambridge University
IBM 3081 computer. Switching between 1% C_3H_6, 2% CO, 3% O_2 and 1% C_3H_6, 4% CO,
3% O_2 is the example considered here.

The simulations are shown in fig.4 to fig.5. Here feed period is 3.0 and
2.5 min, give responses of 6.0 and 5.0 min respectively. This scheme is similar
to experimental scheme X2, in that only the CO concentration is switched, but
it will be recalled that scheme X2 did not lead to any S.H.R. phenomena, whereas scheme X1 (in which the concentration of C_3H_6 was the time-varying variable)
did. This difference may arise from the fact that both reactants are treated
similarly in our reaction model. Recent studies (see (14) for a summary), however indicate that adsorbed CO on Pt aggregates as islands, whereas, so far, no
analogous behaviour has been reported for alkenes. For any given switching
scheme, the feed forcing period can produce oscillatory phenomena of each of
the types identified by Sincic and Bailey (8). The intermediate region, where
pathological phenomena, such as S.H.R. occur are as shown. The region is
bounded by a forcing period of 1.5 minutes and 3.5 minutes. Further it can
exhibit S.H.R. of order 1/2 and 1/3. This kind of response appears to arise
out of harmonic oscillations in which the amplitude of alternating peaks con-
tracts and ultimately disappears into a relaxed steady state response.

CONCLUSIONS

Note that in order for oscillations to appear with a period different from that
of the feed, synchronous reaction rate oscillations are required on all the
catalyst sites in the reactor. This synchronicity, we believe, is provided by

the gas phase above the catalyst. Unless the gas phase is considered to oscillate with the output period, some other means of coupling between surface sites must be invoked, which would make a model of the reaction far more complicated than was intended here. There is little doubt though that, where applicable, dynamic techniques provide a probe which can lead to greater understanding and to the formulation of more precise kinetic models of catalytic processes.

REFERENCES

(1) C.J. Hawkins, Ph.D. thesis, University of Cambridge (1978).

(2) M.B. Cutlip, A.I.Ch.E.J., 25, 502-508 (1979).

(3) M.B. Cutlip, C.N. Kenney, W. Morton, D. Mukesh and S.C. Capsaskis, Proceedings ISCRE-8, I.Chem.E. Symp.Ser., 87 (1984).

(4) K. Tomita and T. Kai, J.Stat.Phys., 21, 65 (1979).

(5) B.L. Hao, J.Theor.Biol., 98, 9-14 (1982).

(6) J.E. Bailey, Chem.Eng.Commun., 1, 111-124 (1973).

(7) J.E. Bailey, in "Chemical Reactor Theory-A Review", L. Lapidus, N.R. Amundson (eds.), Prentice Hall, Englewood Cliffs, NJ, 758-813 (1977).

(8) D. Sincic and J.E. Bailey, Chem.Eng.Sci., 32, 281-286 (1977).

(9) N. Minorsky, 'Non-linear oscillations', Van Nostrand, New York (1962).

(10) D. Mukesh, C.N. Kenney and W. Morton, Chem.Eng.Sci., 38, 69-77 (1983).

(11) S.C. Capsaskis, Ph.D. thesis, University of Cambridge, to be submitted.

(12) M. Sheintuch and D. Luss, Ind.Eng.Chem.Fundam., 22, 209-215 (1983).

(13) M.B. Cutlip, C.J. Hawkins, D. Mukesh and C.N. Kenney, Chem.Eng. Commun., 22, 329-344 (1983).

(14) D. Mukesh, W. Morton, C.N. Kenney and M.B. Cutlip, Surf.Sci., 138, 237-257 (1984).

LIST OF SYMBOLS

c_{TOT}	total gas phase concentration (mol ml^{-1})
k_{p1}, k_{p2}, k_{p4}	CO, O_2 and C_3H_6 adsorption rate, constants (ml mol^{-1} s^{-1})
k_{m1}, k_{m2}, k_{m4}	desorption rate constants (s^{-1})
k_3, k_5	CO, C_3H_6 surface oxidation rate constants (s^{-1})
k_{p6}, k_{p7}	C_3H_6, CO_2 support adsorption rate, constants (ml mol^{-1} s^{-1})
k_{m6}, k_{m7}	C_3H_6, CO_2 support desorption rate, constants (s^{-1})
m	mass of catalyst (gm)
R_i	reaction rate for step i (s^{-1})
s, s'	catalyst active site; support adsorption site
s_o, s_{sup}	total concentration of active sites; of support sites/catalyst mass (mol gm^{-1})
V	reactor volume (ml)
x_i, z_i	gas phase mole fraction and surface conc'n of species i (dimensionless)
z_o	surface capacity factor
τ	period of external forcing
Φ	volume correction term

Subscripts

F	feed
rs	relaxed steady state
species 1	CO
2	oxygen
3	C_3H_6
4	CO_2
v	vacant

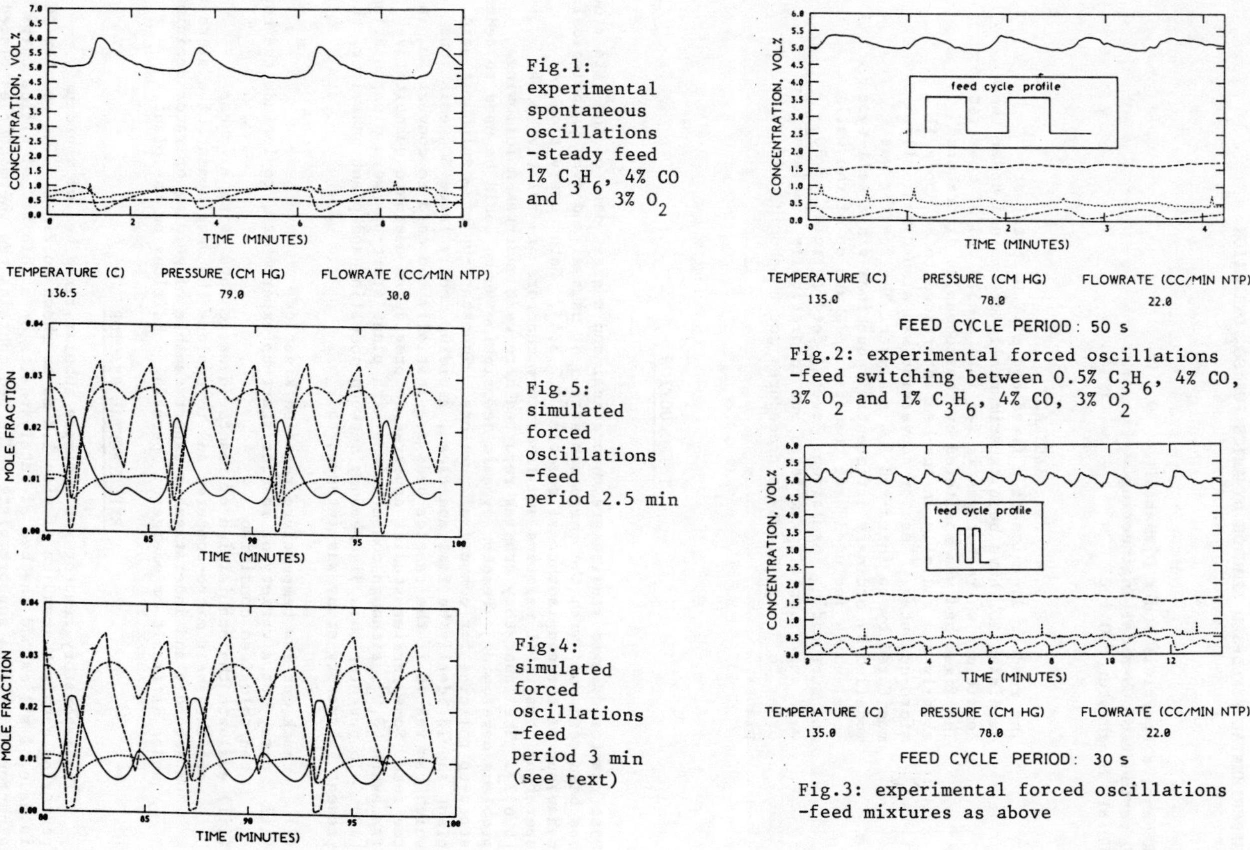

Fig.1: experimental spontaneous oscillations -steady feed 1% C_3H_6, 4% CO and 3% O_2

Fig.5: simulated forced oscillations -feed period 2.5 min

Fig.4: simulated forced oscillations -feed period 3 min (see text)

FEED CYCLE PERIOD: 50 s

Fig.2: experimental forced oscillations -feed switching between 0.5% C_3H_6, 4% CO, 3% O_2 and 1% C_3H_6, 4% CO, 3% O_2

FEED CYCLE PERIOD: 30 s

Fig.3: experimental forced oscillations -feed mixtures as above

EXPERIMENTAL FIXED-BED REACTOR DYNAMICS FOR SO_2 OXIDATION

R Mann, E Stavridis and K Djamarani
Department of Chemical Engineering
UMIST Manchester M60 1QD

ABSTRACT

The reactor dynamics of a fixed-bed of V_2O_5 catalyst
oxidising SO_2 have been experimentally determined under the
conditions where hot gas is fed into a cold bed of catalyst.
The fixed-bed was monitored by thermocouples distributed
axially and radially in the bed interfaced to an Apple II
micro-computer. Reactor conversion was also dynamically
monitored by an infra-red SO_2 analyser. The reactor was
modelled by ordinary differential equations via a cell-type
model which included heat transfer and warm up of the walls
and surrounding insulation. Several feed policies of SO_2
were experimentally explored and theoretically simulated, in
order to devise start-up procedures for sulphuric acid
plants.

INTRODUCTION

Most reactor dynamic studies are theoretical and a great deal of the literature
has been concerned with the characterisation of unusual and even pathological
behaviour of the non-isothermal reactor [1,2,3,4]. Both observations on
operating commercial reactors and laboratory reactors are relatively few
[5,6,7]. The laboratory studies reported here were undertaken following
problems developing a reactor dynamic description which could be used to deduce
start-up policies for commercial reactors. The start-up of a sulphuric acid
plant should ideally be fast and clean, avoiding unduly large SO_2 emissions
which can result if the reactor beds are initially too cold to convert SO_2 in
the feed. Some earlier studies appeared to provide an adequate quantitative
framework [8,9], although comparisions with plant start-ups were disappointing
[10]. In pursuing this discrepancy between modelling and plant behaviour, this
present laboratory study was intended to:

(i)   check out low temperature reaction kinetics
(ii)   carry out a variety of reactor start-up experiments, to include the hot
        bed/cold feed condition
(iii)   evaluate the acquisition of reactor dynamic data using a cheap
        off-the-shelf micro-computer and to explore the development of a micro as
        a friendly and indefatigable expert capable of guiding operator decisions
        under difficult or emergency operating conditions on the plant.

EXPERIMENTAL REACTOR

The laboratory reactor was a simple tubular fixed bed of vanadium
pentoxide catalyst granules of size 1.2 mm as shown in Fig 1. The reactor tube
is made of silica and the bed of catalyst sits on a thin bed of silica chips on
a perforated silica support plate. The bed contains ten thermocouples placed
axially and radially as indicated in Fig 1, with two thermocouples measuring

the feed temperature 3 mm ahead of the catalyst bed itself. The twelve thermo couple leads are threaded through the perforated base plate and led away in the cold downstream section through a liquid mercury seal. Insulation is provided to a thickness of 22.5 mm to restrict heat losses.

The feed stream of air/SO_2 passes through a furnace and reaches a 3-way by-pass valve some 70 mm ahead of the catalyst bed. This valve is operated manually in conjunction with a second valve, so as to cause the feed stream to be switched from by-passing the reactor to flow directly into the fixed bed. The by-pass switchover is virtually instantaneous and the flow rate remains unchanged. In this way, the bed can be kept 'cold' at more or less ambient temperatures and a hot feed stream , with or without SO_2 , can be instantly directed into the reactor. This provides the important experimental cold bed/hot feed condition relevant to plant start-up problems.

During a start-up, sensible heat is transported into the fixed bed in the gas stream flowing from the furnace. To this is added any heat released due to reaction, which is sensitively dependent upon temperature. Thus the temperature distribution in the bed results from heat input from both reaction and the feed stream. The bed temperature transients are monitered by the twelve thermo-couples. The consequent reaction is measured by an infra-red SO_2 analyser downstream following removal of any SO_3.

The thirteen analogue millivolt signals which characterise the reactor dynamics are sampled and read by an Apple II micro-computer using the arrange-ment shown in Fig 2. Regular sampling of these signals through a single channel A/D converter is achieved by a normal multiplexing arrangement, which includes two extra millivolt signals for routine calibration of the A/D card. The Apple II has four annunciator outputs through its games input/output connector. These four outputs can only drive four relays and as a result a demultiplexer (a four to sixteen decoder) had to be used. The four outputs are activated either high or low through a stipulated memory location by simple Basic 'peek' commands which give sequential sampling of the reactor's thirteen analogue signals. Individual A/D conversions take less than 50 µs. With multiple sampling of each thermocouple, the reactor condition can be easily logged at 1 minute intervals using Basic programming without recourse to machine code. In this way, dynamic changes taking place over periods of the order of 30 minutes can be determined with good sensitivity.

STEADY STATE FIXED BED HEAT TRANSFER

The heat transfer characteristics of the fixed bed at steady-state in the absence of any reaction (air feed only) provide the means of estimating the principal heat transfer parameters. Fig 3 shows the axial and radial temp-erature profiles at a feed rate of 7.04 mmol s^{-1} and feed temperature of 410°C. The laboratory reactor bed is clearly non-adiabatic (in spite of the extensive insulation) in contrast to operating plants. Reasonably severe radial and axial temperature profiles are present. Because axial dispersion is negligible [11], the relevant heat transfer parameters are λ_r (the effective radial thermal conductivity) and α_w (the wall heat transfer coefficient). The governing equations without reaction are therefore

$$G C_p \frac{\partial T}{\partial z} = \lambda_r \left(\frac{\partial^2 T}{\partial r^2} + \frac{1}{r} \frac{\partial T}{\partial r} \right) \tag{1}$$

within the catalyst bed, along with the wall condition

$$\alpha_w (T_w - T_r) = \lambda_r \left(\partial T / \partial r |_R \right) \tag{2}$$

These equations have a well known analytical result |12|, but in order to find λ_r and α_w the variation of wall temperature with length $T_w(z)$ must be known.

It was estimated by measurement of the outer surface temperatures of the in
sulation at each of the four axial locations of bed thermocouples, coupled with
solving for radial heat conduction from the reactor wall, through the glass and
insulation into the ambient air. A two-dimensional search then found parameter
values minimising the discrepancy between the ten experimental and theoretical
bed temperatures [12]. The radial Peclet numbers over a range of N_{Re} from 10
to 30 were in good agreement with previous results [13], but the wall Biot
numbers were low in comparison to Dixon and Cresswell [13], though this is
thought to be due to the use of unusually small catalyst particles in the
present study.

Fig 4 shows the reasonable correspondence between experimental and
theoretical profiles in a typical measurement. Theory tends to overpredict
temperatures in the entry part of the bed and to underpredict towards the exit.
Agreement is good in the central part of the bed. The discrepancies probably
reflect axial conduction effects at the ends of the fixed bed.

<u>THERMAL DYNAMICS WITHOUT REACTION</u>

In order to avoid computations with PDE's, the catalytic reactor is considered
to approximate plug flow by the use of 12 backmixed stages. To incorporate the
dynamics of the glass wall and the insulation, as well as external heat
transfer from the insulation to the ambient air, these are each considered to
form hypothetical 'backmixed' lumps comprising the same number of axial stages,
with negligible axial heat transfer. Such a so-called cell model approach as
illustrated in Fig 5 limits the numerical simulations to the simple solution of
sets of ordinary differential equations forming an initial value problem. Many
previous approaches to analysing fixed-bed dynamics have failed to take
reasonable account of the thermal dynamics of reactor accoutrements. This
lumped cell-model provides an approximation to the radial profiles as indicated
in Fig 6. The appropriate value for the overall heat transfer coefficient for
each backmixed stage is given by

$$\frac{1}{U} = \frac{1}{\alpha_w} + \frac{D}{8\lambda_r} \tag{3}$$

The simulated unsteady heating up of the fixed bed is shown in Fig 7 in
terms of the mean axial temperatures. The initial reproduction up to 30
minutes is good, although the final steady-state axial profile predicts
somewhat higher theoretical temperatures than are experimentally observed,
reflecting of course the same shortcomings present in Fig 4, which seem to be
related to end effects. The predicted outer surface temperatures of the
insulation as the reactor comes on stream are shown in Fig 8 with a temperature
profile for each of four minute intervals. The final steady state is closely
approached after 100 minutes. Fig 9 then shows how the radial temperature
profiles evolve in the first stage after hot feed is directed into the reactor.

<u>FIXED-BED DYNAMICS WITH REACTION</u>

Four different experimental feed policies for SO_2 after switching the flow
through the reactor were explored. These are shown in Fig 10, and comprise
immediate addition of SO_2 at the design rate of 10% (by volume), delayed
addition with a step to the full rate of 10%, and two four increment step
policies based on five and eight minute time intervals. The immediate step to
the design rate of 10% gives the sxperimental reactor dynamics shown in Fig 11,
where they are compared with theoretical simulations.

For reaction, the theoretical model uses the Mars-Maessen | 14 | equation:-

$$-r_{SO_2} = k\, P_{O_2} \frac{K_M\, P_{SO_2}/P_{SO_3}}{\left[1 + K_M (P_{SO_2}/P_{SO_3})^{1/2}\right]} \left[1 - \left(\frac{P_{SO_2}}{P_{SO_2}\, P_{O_2}^{1/2}\, K_p}\right)^2\right]$$

with k = 4.47 x 10^7 exp (-28.3/RT)   for T > 650°K

  k = 1.00 x 10^3 exp (-45.8/RT)   for T < 650°K

Effectiveness factors were close to unity up to 750°K because of the small size of the catalyst particles. In Fig 11, the reactor warm-up is shown to correspond quite closely to the dynamic model, with the greatest deviations being evident in the front of the fixed bed, probably reflecting an end effect due to axial conduction. A comparison of the measured and predicted dynamic conversion of SO_2 is shown in Fig 12. The predicted conversions are slightly lower on average than those experimentally measured, but overall the timewise increase in conversion towards the steady-state value of 79% is closely reflected by the theoretical model. This is also the case for each of the other three experimental SO_2 feed policies shown in Fig 12. The theoretical predictions for each experimental feed policy are compared together in Fig 13, from which the superiority of the four-step policy implemented at eight minute intervals is quite clear.

TOWARDS AN OPTIMAL START-UP POLICY

The fact that the theoretical reactor modelling can satisfactorily reproduce experimental reactor behaviour under a variety of feed policies of SO_2 suggests that the theory may be reasonably used to seek out an optimal policy providing a fast non-polluting start-up. In this respect non-polluting means that the mass flow rate of SO_2 leaving the reactor does not exceed the design steady-state value. Fig 14 then shows the mass flow rate of SO_2 relative to a design value of unity for two ramped increases of feed SO_2 and an intermediate stepped policy. In each case the reactor effluent is predicted to exceed the permitted level of emission for significant periods, although at first in each case the emission undershoots.

It is a simple matter to iterate on the feed profile of SO_2 so as to bring the reactor on stream without exceeding the 'statutory' level. Fig 15 shows a close approximation to a perfect non-polluting start-up of the laboratory reactor from cold, which uses 4% of SO_2 in the feed over an initial period of 40 minutes, followed by a ramp up to the design value of 10% over a final 10 minute period. The experimental verification of reactor dynamic modelling of the laboratory reactor in the present study confirms that this approach should be able to be successfully applied to deduce fast clean start-up procedures for the more demanding requirements of commercial plants.

REFERENCES

[1]   Heineman, R.F. and Poore A.B., Chem.Eng.Sci., 36,1411,1981

[2]   Morbidelli, M. and Varma, A., A.I.Ch.E Jl.,28, 705, 1982

[3]   Akella, L.M. and Lee, H., A.I.Ch.E.Jl, 29, 87, 1983

[4]   Jorgensen D.V. and Aris, R., Chem.Eng.Sci, 38,45,1983

[5]   Sharma C.S. and Hughes, R., Chem Eng Sci, 34, 613, 1979

[6]   Kalthoff, O. and Vortmeyer, D.,Chem EngSci, 35, 1637, 1980

[7]   Bonvin, D., Rinker R.G and Mellichamp, D.A, Chem Eng Sci, 38,233,1983

[8]   Mann, R., Gardner I.J., and Morris,C.,I.Chem.E.Symp.Series 57 LI, 1979

[9]   Mann, R, Gardner I.J. and Morris, C., Chem.Eng.Sci,35,185,1980

[10] Gardner I.J., PhD Thesis, UMIST 1980

[11] Mecklenburgh, J.C., Trans.I.Chem.Engrs.,52, 180,1974

[12] De Wasch A.P. and Froment, G.F., Chem.Eng.Sci., 27,567,1972

[13] Dixon, A.G. and Cresswell, D.L., A.I.Ch.E.J., 25,663,1979

[14] Mars P.,and Massen,J., Jnl.Catalysis, 10,1,1968

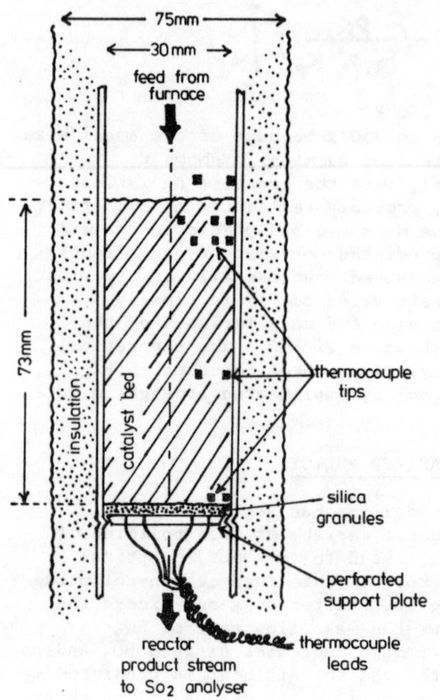

FIG 1 Experimental fixed-bed reactor

FIG 2 The Computer-Rig Interface

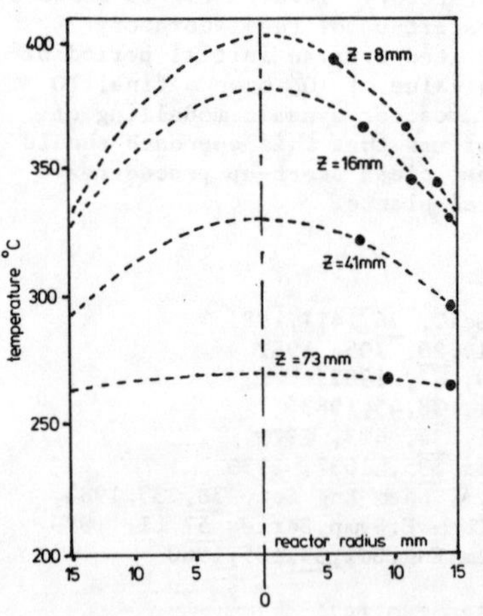

FIG 3 Temperature Profiles Without Reaction

FIG 4 Theoretical Temperature Profiles

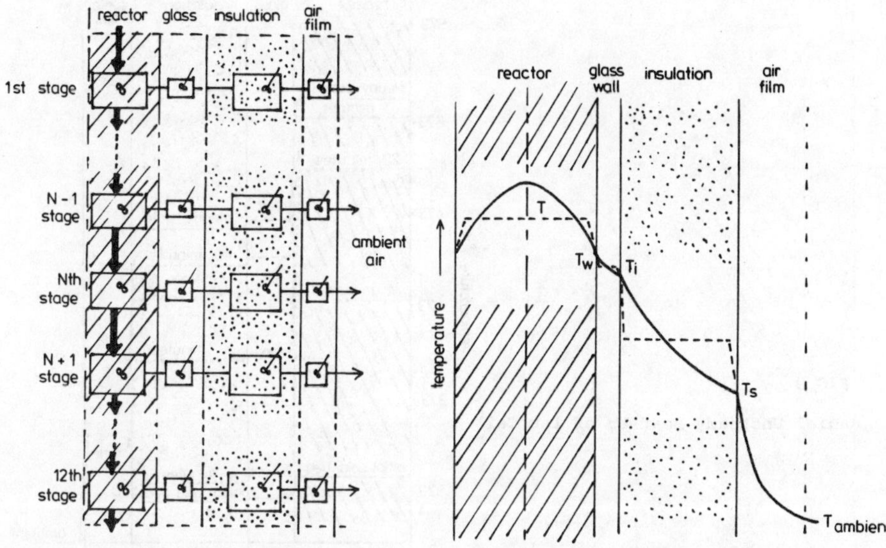

FIG 5 Cell–model for Reactor

FIG 6   Radial Temperature Profiles

using a Cell Model

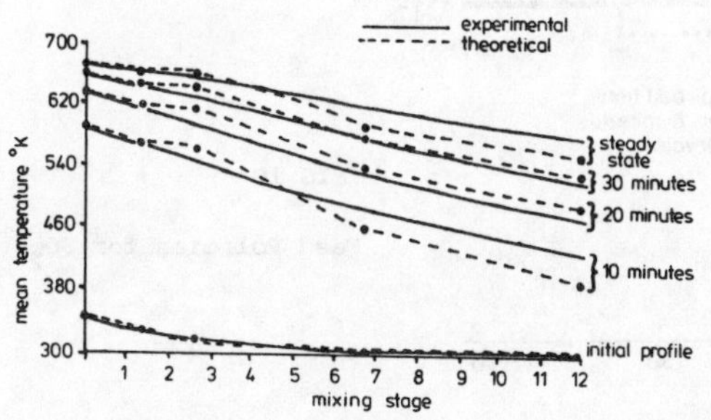

FIG 7 Mean Axial Temperatures Without Reaction

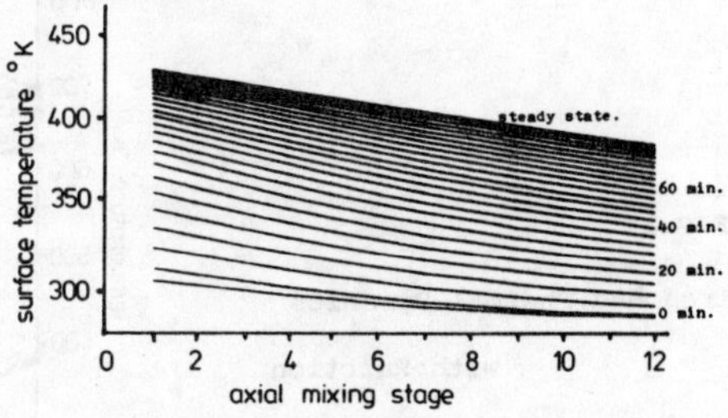

FIG 8 Dynamic Heat–Up of Insulation Surface

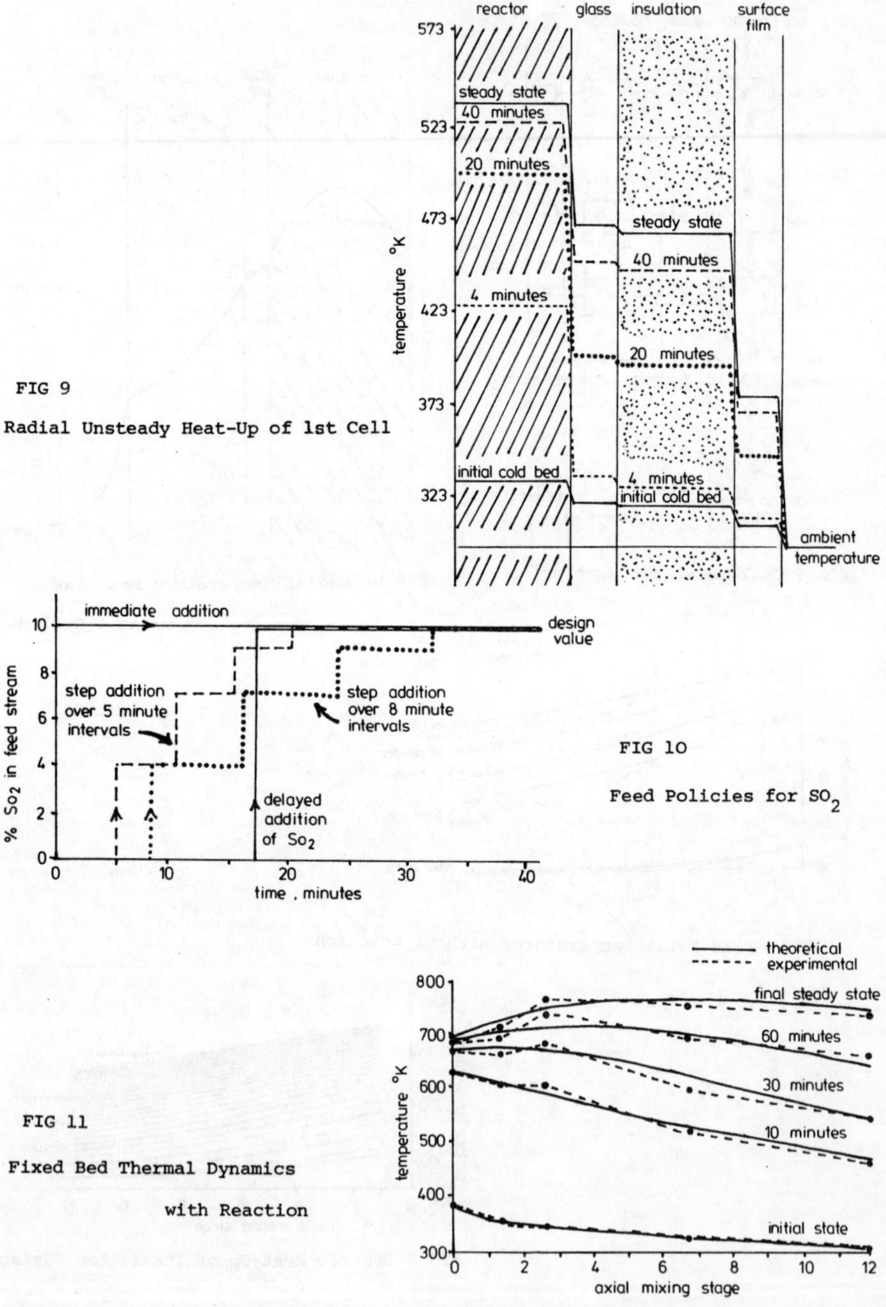

FIG 9

Radial Unsteady Heat-Up of 1st Cell

FIG 10

Feed Policies for SO$_2$

FIG 11

Fixed Bed Thermal Dynamics

with Reaction

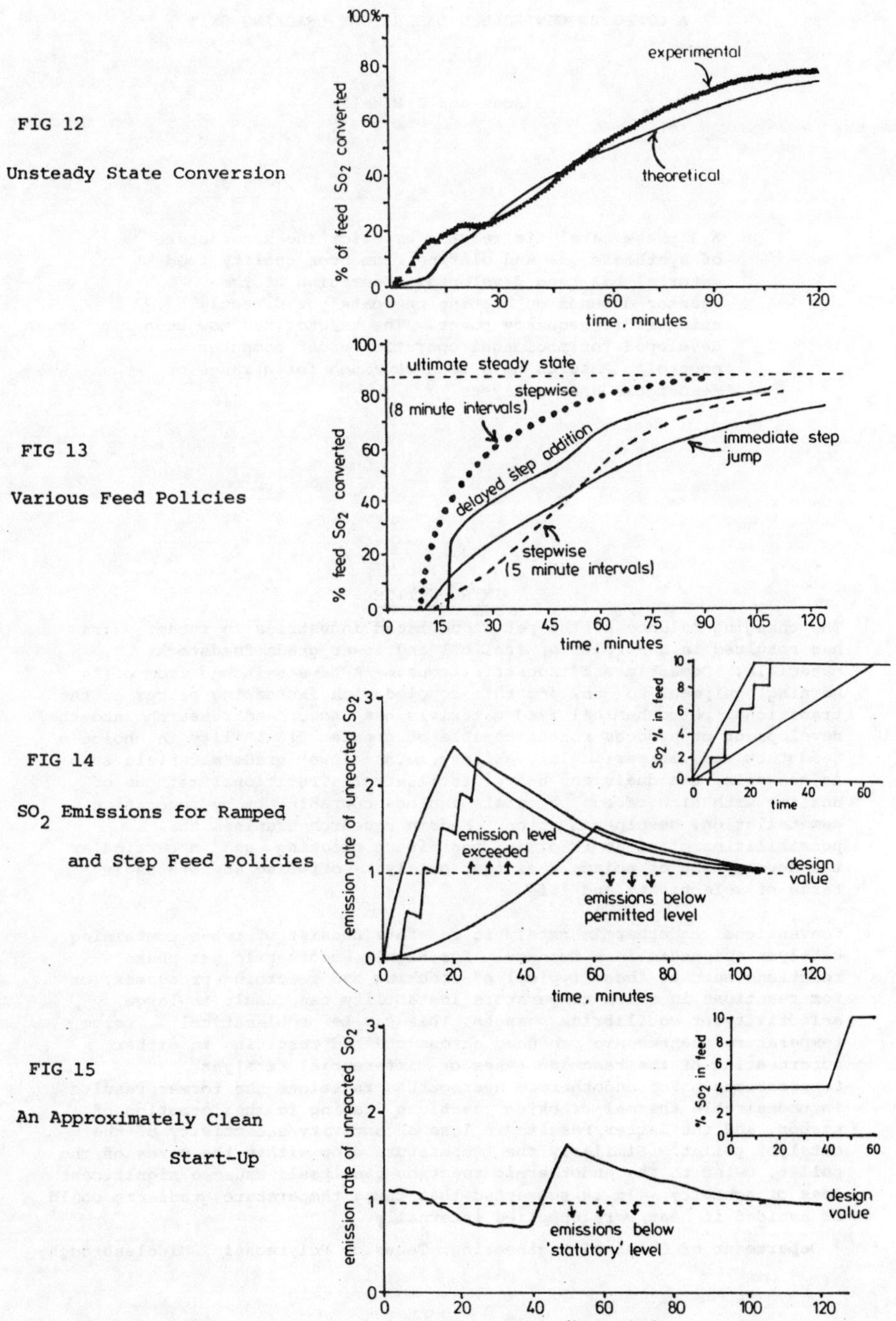

FIG 12

Unsteady State Conversion

FIG 13

Various Feed Policies

FIG 14

SO$_2$ Emissions for Ramped and Step Feed Policies

FIG 15

An Approximately Clean Start-Up

A COMPUTER CONTROLLED LABORATORY CRACKING UNIT

S Acey and J R Walls[*]

A 3 phase catalytic reactor used for the manufacture
of synthesis gas and olefins from poor quality feed
material has been developed. Operation of the
reactor depends on heating the catalyst directly
using radiofrequency power. The reactor has now been
developed for prolonged operation under computer
control. Results have been obtained for a range of
feedstocks and catalysts.

INTRODUCTION

The changing balance of the petroleum based industries in recent years
has resulted in a surplus of fuel oil and lower grade feedstock
materials. Domestic and industrial consumers have switched from oil
burning equipment to gas, and this coupled with increasing prices of the
traditional petrochemical feed materials has encouraged research into the
development of process routes capable of greater flexibility in choice of
feedstock and, in particular, ability to use lower grade materials such
as gas oils, residuals and heavy distillates. Traditional methods of
dealing with high carbon materials include upgrading by hydrogenation,
demetallation, desulphurisation. Modern research examines the
possibilities of steam reforming and direct cracking, and in particular
the development of poison resistant catalysts offering advantages in
terms of selectivity and life.

Conventional endothermic catalytic reactors consist of tubes containing
catalyst suspended in a furnace. For highly endothermic gas phase
reactions such as those typical of cracking and reforming processes, or
for reactions in which temperature instability can result in large
selectivity or equilibrium changes, this can be problematical . Large
temperature changes are produced across the bed resulting in either
superheating of the reacting gases or differential catalyst
temperatures. For endothermic hydrocarbon reactions the former results
in undesirable thermal cracking reactions leading to the formation of
carbon, and the latter results in loss of activity/selectivity of the
catalyst pellet. Similarly the temperature drop within the pores of the
pellet, owing to the endothermic reaction can itself cause a significant
loss of activity. It is suggested that these temperature gradients could
be avoided if heat were supplied internally.

* Department of Chemical Engineering, Teesside Polytechnic, Middlesbrough

THE HEATING TECHNIQUE

An alternative technique has been developed by NITHIANANTHAN and
WALLS (1). By suspending a suitable material in an electromagnetic
radiofrequency field it is possible to heat that material directly. In
this way the endothermic heat requirement can be supplied from within the
catalyst pellet, without heating of either reaction materials or vessel.
Radiofrequency heating may be broadly split into two categories:
induction heating and dielectric heating. An electromagnetic wave
consists of two perpendicular components, the electric and magnetic field
vectors. Induction heating is essentially concerned with the heating of
conductors utilising the magnetic field, and is generally performed at
lower frequencies, whereas dielectric heating utilises the electric field
vector for heating and occurs at higher frequencies (up to 100MHz.).
Catalysts consist of a complex mix of conducting and non-conducting
materials, the heating mechanism being correspondingly complex is
discussed by OVENSTON and WALLS (2, 3). Advantages offered by this
method of catalyst heating include

(i)   heating from within leads to a more even temperature distribution
        across the catalyst, which in turn leads to a reduction of coking
        and sintering at hot spots, as well as a more consistent product
        composition.

(ii)   heat of reaction is provided from within the pellet so removing
        the need to preheat feed materials, and minimising pyrolysis in
        the bulk fluid.

(iii)   more accurate and rapid response is achieved as a result of the
        direct heating technique and lower thermal capacity.

Initial work was performed on vapour phase steam reforming of naphthas,
examining thermodynamic equilibrium compositions and steam/carbon ratios
necessary to avoid coking of the catalyst. Although such a system has
been shown to operate successfully it is less suitable for use with heavy
oils and residues. Problems are encountered vaporising such materials,
fractions of which may boil at temperatures in excess of 500°C. Heating
to this temperature results in carbon formation problems within the
vaporising unit and in the gas phase. An alternative form of 3 phase
reaction system has been developed. A mixture of oil and steam is fed to
the reactor. Mixing occurs below the catalyst, and reaction occurs on
the catalyst which is surrounded by a pool of boiling oil. Energy
required to heat the feedstock to its boiling point and vaporise it is
generated within, and transferred from the catalyst surface. It is
proposed that reaction takes place in a vapour film surrounding the
catalyst, the boiling pool of oil providing a natural and immediate
quench which prevents the attainment of equilibrium which would degrade
the feedstock to methane, hydrogen and carbon at the temperatures
typically used (800°C). Additionally catalytic action assists the
formation of CO, CO_2 in preference to carbon. The reaction may be
thought of as a mixture of reforming and cracking reactions.

$$C_nH_m + nH_2O \longrightarrow nCO + (n + m/2)H_2 \tag{1}$$

REFORMING

$$CO + H_2O \longrightarrow CO_2 + H_2 \tag{2}$$

$$C_nH_{2n} +2 \longrightarrow C_mH_{2m} + C_pH_{2p} +2 \qquad (n = m + p) \qquad (3)$$

paraffin $\qquad$ olefin $\quad$ paraffin

CRACKING

$$C_nH_{2n} \longrightarrow C_mH_{2m} + C_pH_{2p} \qquad (n = m + p) \qquad (4)$$

olefin $\qquad$ olefin $\quad$ olefin

Major reaction products formed are synthesis gas, ethylene, propylene and carbon dioxide. Under these conditions the reactor system was found to operate successfully for several hours without carbon formation on the catalyst. Occurrence of carbon formation is easily noticed. Carbon is a good conductor, and, as such rapidly heats to temperatures in excess of 1000°C, and is evidenced by bright yellow or white spots as carbon burns off the catalyst surface.

THE COMPUTER FACILITY

Further evaluation of the process was required. It was decided that this could best be achieved using a computer operated control and monitoring system. The operation of the process is based around the Departmental Ferranti Argus 700GL computer. The computer is dedicated to process control, and is capable of simultaneous operation of up to 12 laboratory/pilot scale plants. The control system was designed and constructed on the basis of developing a computer controlled laboratory reactor which would:

(i)   offer flexibility in the choice of start-up, operating and data recording procedures.

(ii)   provide an automatic, reliable and accurate data monitoring facility enabling a more extensive analysis of reactor performance and catalyst evaluation.

(iii)   provide a control facility allowing the process to be safely operated without supervision.

(iv)   be capable of a controlled shutdown procedure in the event of an

(v)   allow for development of a reactor suitable for both computer and manual operation.

The Ferranti Process Management System is configured in such a way that its use as a tool by the Chemical Engineer requires no advanced knowledge of real time programming languages, but allows the engineer to construct and implement his own software. This is achieved by offering a set of fully tried and tested software packages written in the programming language CORAL. These packages can be linked together to provide the basis of a data logging and control facility. Additional requirements outside the range of pre-programmed packages are available by means of user written CORAL and FORTRAN programs. Plant signals are interfaced to the computer as analogue and digital input and output signals whose operation is governed by control loops, into which the appropriate PID parameters may be entered, and control and data logging sequences for startup, operation and shutdown.

THE PROCESS

A process flowsheet is shown in Fig 1. Oil and water are fed to the reactor by means of peristaltic pumps. Steam is generated in an electrical unit. Steam and oil are contacted in a glass reaction tube below the heated catalyst. Reaction takes place at the catalyst surface in a 3 phase system. All the feedstock is eventually vaporised. The catalyst pellets are suspended in the centre of the radiofrequency heating coil. The power rating for the R.F generator is 3kW., and the operating frequency is 20.6MHz. Temperature is measured using an infra-red optical fibre probe and controlled by altering the setting of a 3 phase variac controlling power input to the generator. The use of optical fibres ensures the transmission of an interference free signal from within the R.F field. It was anticipated that in a high noise radiofrequency environment problems would be encountered with signal interference. However it was found that by using screened cable and wherever possible running signals outside a wire mesh Faraday cage these problems were avoided. Only where there was direct contact with the radiofrequency field (ie temperature measurement) was optical isolation necessary.

After reaction the vapour mixture consisting of reaction products and unreacted steam and recycle oil is passed through two water cooled condensers into a catchpot. From the catchpot liquid product passes to a continuously operating digital balance, while vapour product passes through a thermal massflowmeter to the gas chromatograph before being burnt within a flare. Gas analysis is performed on a PERKIN ELMER F30 gas chromatograph fitted with an automatic gas sampling valve. Temperature programmed analysis by both a hot wire detector and a flame ionization detector is possible. This allows gas sampling at 30 min. intervals. Automatic operation of the various devices and detailed calculation is controlled by a BASIC program actuated from a PERKIN ELMER SIGMA 15 datastation. Analysis allows examination of the C1 to C4 material in the product gas.

Data logged by the Ferranti computer are recorded at 5 min. intervals, an average value being recorded where appropriate. A datalink enables transfer to the Polytechnic PRIME 9750 computer for detailed calculation. Additional sequences created allow the continuous recording and plotting of parameters at shorter intervals if required.

RESULTS

Extensive results have been obtained using a range of heavy industrial feedstocks (4). The current work concentrates predominantly on two feedstocks, Kuwait Wax distillate, consisting of material boiling between 268 and 517°C, and n-dodecane. Wax Distillate has proved to be the most satisfactory of the heavy feedstocks tried, and has been used to examine properties of the system, and to compare various catalytic materials. n-dodecane allowed the examination of the liquid effluent from the reactor, something which would have proved impossible with the complex feedstocks previously used. Table 1 gives typical results for operating runs with these feedstocks and North Sea Crude Oil. The advantages of being able to directly crack a Crude Oil are evident, however the wide boiling range of material present in the feed prevented the establishment of a stable boiling regime, resulting in frequent "flooding" of the

reactor and inability to successfully operate at high temperatures.
Reaction products depend significantly on temperature which proves to be
the major parameter affecting product composition. At lower temperatures
the major products are synthesis gas and carbon dioxide. As temperature
increases a greater proportion of cracking occurs resulting in the
formation of olefinic materials, and in particular ethylene and
propylene. Product distributions are given for a range of different
materials and catalysts in Table 1. Initial work was performed using a
commercial nickel on silica/alumina catalyst(A). Such a catalyst showed
useful results, but other materials giving better heating performance in
the radiofrequency field were sought. The most successful of these was
Fe_3O_4 (C1-C10). Additionally results were recorded for commercial Fe_3O_4
based pellets(B). A pellet containing 40% Fe_3O_4 w/w can be maintained at
reaction temperature using considerably less power. Over a period of
several hours no noticeable loss of activity was found for the nickel
pellets, but the Fe_3O_4 pellets slowly decay in activity, resulting in
carbon formation and destruction of the catalyst pellet. Experiments
with n-dodecane as feedstock revealed the importance of the boiling pool
of oil surrounding the catalyst. It has been stressed that the
combination of catalyst and quench is necessary to prevent carbon
formation. When using n-dodecane which has a boiling point of 217°C the
vapour film around the catalyst expands to fill the entire tube, causing
the reactor to operate as a gas phase reactor. Under such conditions
carbon rapidly forms on the catalyst.

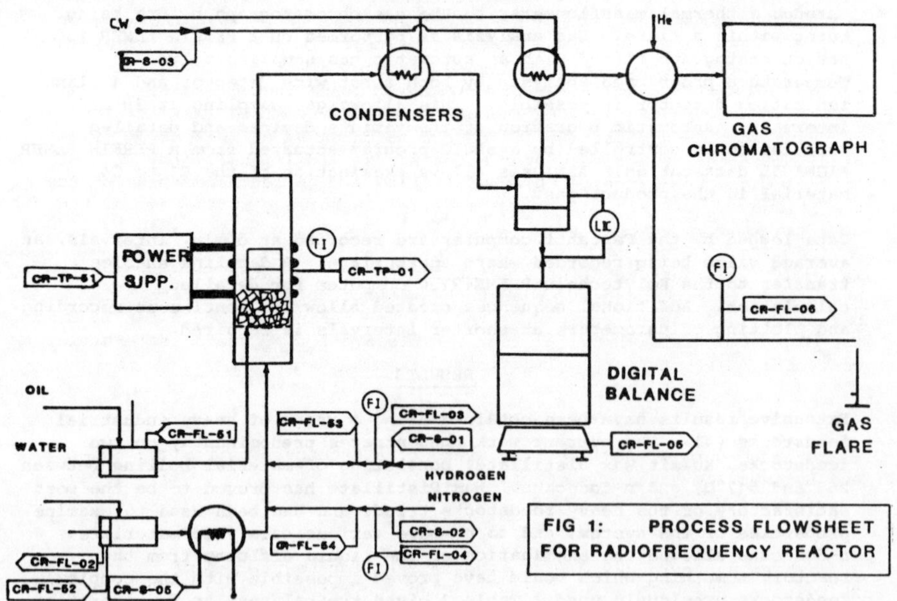

TABLE 1 Product Distribution for Various Feedstocks

FEEDSTOCK	KUWAIT WAX DISTILLATE				n-DODECANE			NSC*
CATALYST	C1	A	A	B	A	C4	C4	A
RUN	17-2	40-5	41-4	48-1	29-1	30-1	31-2	9-2
WT% H_2	2.81	4.53	14.72	3.15	2.78	1.40	1.04	0.48
CO	16.13	6.41	5.33	20.66	10.76	48.20	28.91	22.60
CH_4	2.62	3.05	3.73	4.56	2.86	4.07	5.40	0.00
CO_2	42.41	30.91	38.92	33.15	83.59	30.51	27.81	36.45
C_2H_4	5.91	11.23	13.11	11.76		8.74	21.29	1.04
C_2H_6	1.71	1.31	1.59	2.81		1.49	1.63	1.20
H_2S	0.37	1.26	0.18	5.07				0.46
C_3	1.51	9.54	9.25	9.36		3.77	7.82	12.75
C_4	6.10	12.46	10.77	9.49		1.82	5.16	10.72
$>C_4$	10.53	19.72	12.40				0.94	14.31
LHSV /hr	13.0	6.1	10.2	13.3	11.2	15.3	11.2	2.0
TEMP deg.c	671	678	697	652	613	700	800	578
GASN wt%	4.8	19.5	10.1	8.2	7.0	1.5	6.2	20.4
S/C ratio	1.3	3.1	3.4	2.1	3.3	3.2	3.2	4.9

* North Sea Crude Oil

CONCLUSIONS

The results demonstrated here have been obtained in the final stages of commissioning of the computer operated system. It should now be possible to operate the reactor over prolonged periods investigating changes in product composition and catalyst performance with respect to time. In experimental work to date the average level of feedstock gasification has not been seen to deteriorate, which can be taken as an indication of a negligible loss of activity over a few hours. Additionally the accumulation of a large database will allow a detailed assessment of the conditions required to create the stable film boiling regime around the catalyst.

In summary the unit developed, which can now be operated by computer has been shown to process heavy feedstocks without carbon formation. In principle there is no reason why the heating technique should not be used for other reactions where the requirements are for accurate temperature control and direct heat application.

ACKNOWLEDGEMENTS

S Acey wishes to acknowledge the award of a studentship from the Science and Engineering Research Council.

REFERENCES

1 Nithiananthan V. 1978 PhD. Thesis Teesside Polytechnic.
2 Ovenston A. Walls J. R. 1983 J Chem Soc Faraday Trans I 79 1073-1084
3 Ovenson A. Walls J. R. 1984 J Phys D Appl Phys 17 L101-L103
4 Walls J. R. Lee J. K. Ovenston A. 1984 I Chem E Symp Ser 87 463-469